This guide is for any[one who has]
viewed a tractor with simple questions.
What is it? How much does it weigh?
How many were built?

I would like to thank the following people
for assisting and providing information
found in this book: Ken Updike,
David Atley, and the Aerostart Angels—
providing hope to every tractor that it
might work again.

**ALL SERIAL NUMBERS AND DATA IN THIS
BOOK HAS ORGINATED FROM OFFICIAL IH
RECORDS**

Tractor Model: AW-6

Rated Horsepower: Belt 32.7   **Drawbar** 25.5

Australian Tractor Test #

Engine: AC-248

Average Shipping Weight: 5,500 lbs (2,500 kgs)

Serial Number Data

Prefix 'WBK' -same as American Built tractors

Units Built	Financial Year (July 1 to June 30)
266	1947/1948
738	1948/1949
1316	1949/1950
1951	1950/1951
2250	1951/1952
888	1952/1953

Tractor Model: Super AW-6

Rated Horsepower **Belt** 45

Australian Tractor Test #

Engine AC-264 Kerosene

Average Weight: Operating 7,068 lbs. (3,206 kg)

Serial Number Data
Prefix AW6

Units Built	Financial Year (July 1 to June 30)
1137	1953/1954
663	1954/1955
250	1955/1956
148	1956/1957
59	1957/1958

Tractor Model: Super AWD-6

Rated Horsepower **Belt** 47 **Drawbar** 42

Australian Tractor Test #

Engine AD-264

Average Weight: Operating: 7,298 lbs (3,310kg)

Serial Number Data

Units Built	Financial Year (July 1 to June 30)
745	1953/1954
1255	1954/1955
924	1955/1956
679	1956/1957
65	1957/1958

Tractor Model: Farmall M—Australian

Rated Horsepower   **Belt** 32.8 **Drawbar** 25.9

Australian Tractor Test #

Engine AC-248

Average Shipping Weight: 5,500 lbs (2,500 kg)

Serial Number Data
Prefix FBK—Same as American tractor

Units Built	Financial Year (July 1 to June 30)
200	1948/1949
700	1949/1950
1,000	1950/1951
1,001	1951/1952
440	1952/1953

Tractor Model: Farmall Super AM

Rated Horsepower   **Belt** 41 **Drawbar** 36.7

Australian Tractor Test #

Engine AC-264

Average Weight: Operating with one set of
weights 7,462 lbs (3,385 kg)

Serial Number Data
Prefix 'AM'

Units Built	Financial Year (July 1 to June 30)
130	1953/1954
82	1954/1955
51	1955/1956
1	1956/1957
0	1957/1958
1	1958/1959

Tractor Model: Farmall Super AM-D

Rated Horsepower   **Belt** 46.5 **Drawbar** 42

Australian Tractor Test #

Engine AD-264

Average Weight: Operating: 6,804 lbs (3,086 kg)

Serial Number Data

Units Built	Financial Year (July 1 to June 30)
120	1953/1954
200	1954/1955
92	1955/1956
52	1956/1957
1	1957/1958
7	1958/1959

Tractor Model: AOS-6

Rated Horsepower **Belt** 37 **Drawbar** 32.5

Australian Tractor Test #

Engine AC-248

Average Shipping Weight: 5,162 lbs (2,342 kg)

Serial Number Data
Prefix AOS6

Units Built	Financial Year (July 1 to June 30)
6	1952/1953
868	1953/1954
800	1954/1955
150	1955/1956
75	1956/1957
7	1957/1958
14	1958/1959

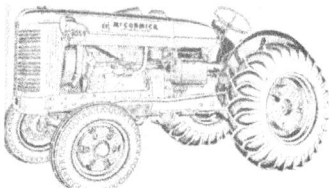

Tractor Model: AW-7 (Kerosene)

Rated Horsepower   **Belt** 41 **Drawbar** 36.7

Australian Tractor Test #

Engine AC-264

Average Shipping Weight: 5,538lbs (2,512 kg)

Serial Number Data

Units Built	Financial Year (July 1 to June 30)
1	1956/1957
46	1957/1958
140	1958/1959
Unknown	1959/1960
Unknown	1960/1961

Tractor Model: AW-7 (Diesel)

Rated Horsepower   **Belt** 48.8 **Drawbar** 42

Australian Tractor Test # 30

Engine AD-264

Average Shipping Weight: 5,768 lbs (2,616 kg)

Serial Number Data

Units Built	Financial Year (July 1 to June 30)
106	1956/1957
871	1957/1958
1,244	1958/1959
Unknown	1959/1960
Unknown	1960/1961

Tractor Model: Farmall AM-7 (Kerosene)

Rated Horsepower   Brake 45

Australian Tractor Test #

Engine AC-264

Average Shipping Weight: 5,650lbs (2,563 kg)

Serial Number Data

Units Built	Financial Year (July 1 to June 30)
8	1957/1958
7	1958/1959
Unknown	1959/1960

Tractor Model: Farmall AM-7 (Diesel)

Rated Horsepower   **Brake** 50

Australian Tractor Test #

Engine AD-264

Average Shipping Weight: 5,650 lbs (2,563 kg)

Serial Number Data

Units Built	Financial Year (July 1 to June 30)
78	1957/1958
65	1958/1959
Unknown	1959/1960

Tractor Model: A-554 (Diesel)

Rated Horsepower   **Belt** 51.7 **Drawbar** 50.8

Australian Tractor Test # 36

Engine AD-264

Average Shipping Weight: 7,280lbs (3,303 kg)

Serial Number Data

501 to 1765	Built in 1961
1766 to 3582	Built in 1962
3583 to 5589	Built in 1963
5590 to 7564	Built in 1964
7565 to 8971	Built in 1965
8972 to 9303	Built in 1966

Tractor Model: A-554 (Kerosene)

Rated Horsepower   **Belt** 52

Australian Tractor Test #

Engine AC-264

Average Weight: Operating: 7,930lbs (3,598 kg)

Serial Number Data

See A-554 list.
Kerosene Model Built from 1961 to 1963

Tractor Model: A-554 Farmall

Rated Horsepower   **Belt** 51.7   **PTO**

Australian Tractor Test #

Engine AD-264

Average Shipping Weight: 7,280lbs (3,303 kg)

Serial Number Data

See A-554 list.
Farmall Model Built from 1964 to 1966

Tractor Model: A-514 Farmall

Rated Horsepower   **Belt** 48.8 **Drawbar** 44.6

Australian Tractor Test #

Engine AD-264

Average Shipping Weight: 7,280lbs (3,303 kg)

Serial Number Data

Built from 1961 to 1964

Unknown with the following exception

150 built in 1962

Highest serial number viewed: 879 (379th built)

Tractor Model: A-414

Rated Horsepower   **Belt** 36   **PTO** 36.7

Australian Tractor Test #

Engine AD-154

Average Weight: Operating: 3,890 lbs (1,765kg)

Serial Number Data

501 to 920   Built in 1963
921 to 3210   Built in 1964
3211 to 4986   Built in 1965
4987 to 5914   Built in 1966

Tractor Model: 564

Rated Horsepower   **Belt** 51.7 **Drawbar** 50.8

Australian Tractor Test # 46

Engine AD-264

Average Shipping Weight: 7,280lbs (3,303 kg)

Serial Number Data

501 to 849	Built in 1966
850 to 1814	Built in 1967
1815 to 2194	Built in 1968
2195 to 2557	Built in 1969
2558 to 2666	Built in 1970
2667 to 2824	Built in 1971
2825 to 2864	Built in 1972

B Series introduced in 1970

Tractor Model: 564 Farmall

Rated Horsepower   Belt 51.7 **Drawbar** 50.8
Australian Tractor Test #

Engine AD-264

Average Shipping Weight: 7,280lbs (3,303 kg)

Serial Number Data

See 564 List

Tractor Model: 564 B Series

Rated Horsepower   **Belt** 51.7   **Drawbar** 50.8

Australian Tractor Test #

Engine AD-264

Average Shipping Weight: 7,280lbs (3,303 kg)

Serial Number Data

See 564 List. Introduced in 1970

Tractor Model: 434

Rated Horsepower   **Belt** 36.6 **Drawbar** 35

Australian Tractor Test # 47

Engine AD-154

Average Shipping Weight: 4,100 lbs (1,860kg)

Serial Number Data

501 to 600	Built in 1966
601 to 1970	Built in 1967
1971 to 2901	Built in 1968
2902 to 3876	Built in 1969
3877 to 4746	Built in 1970
4747 to 5712	Built in 1971
5713 to 5946	Built in 1972

Tractor Model: 686

Rated Horsepower   **Belt** 68 **Drawbar** 61

Australian Tractor Test # 80

Engine D-310

Average Shipping Weight: 9,268lbs (4,204kg)

Serial Number Data

501 to 699	Built in 1969
700 to 1105	Built in 1970
1106 to 1330	Built in 1971

Built with 756. No distinction between models.

Tractor Model: 756

Rated Horsepower   **Belt** 76 **Drawbar** 69

Australian Tractor Test # Nebraska Tractor Test
                                              #955 used.

Engine D-310

Average Shipping Weight: 9,390lbs (4,260kg)

Serial Number Data

See 686 list.

Tractor Model: 696

Rated Horsepower **Belt** 71 **Drawbar** 63.2

Australian Tractor Test # (Used 686 A.T.T. #70)

Engine D-310

Average Shipping Weight: 9,630lbs (4,368kg)

Serial Number Data

501 to 1139	Built in 1972
1140 to 1810	Built in 1973
1811 to 2552	Built in 1974
2553 to 3269	Built in 1975
3270 to 3998	Built in 1976
3999 to 4603	Built in 1977
4604 to 4771	Built in 1978

Tractor Model: 766

Rated Horsepower   **Belt** 78 **Drawbar** 67.5

Australian Tractor Test # Nebraska 955 applies

Engine D-310

Average Shipping Weight: 9,860lbs (4,472kg)

Serial Number Data

See 696 List.

Tractor Model: 444

Rated Horsepower: Max 42 **Drawbar** 35

Australian Tractor Test # 79

Engine AD–154

Average Weight: Operating: 4,840lbs (2,195kg)

Serial Number Data

501 to 790	Built in 1973
791 to 1150	Built in 1974
1151 to 1520	Built in 1975
1521 to 1620	Built in 1976

Tractor Model: 866

Rated Horsepower   **Belt** 96 **Drawbar** 92

Australian Tractor Test # 86

Engine D-358

Average Shipping Weight: 10,790lbs (4,900kg)

Serial Number Data

501 to 600   Built in 1973
601 to 760   Built in 1974
761 to 1044   Built in 1975
1045 to 1566   Built in 1976
1567 to 1997   Built in 1977
1998 to 2142   Built in 1978

Tractor Model: 976

Rated Horsepower Max: 134

Australian Tractor Test #

Engine DT-358

Average Shipping Weight: 11,000lbs (5,000kg)

Serial Number Data

501 to 766	Built in 1976
767 to 1080	Built in 1977
1081 to 1215	Built in 1978

Tractor Model: 786

Rated Horsepower **Max** 100

Australian Tractor Test #

Engine D-310

Average Shipping Weight: 10,350lbs (4,705kg)

Serial Number Data

501 to 720   Built in 1978
721 to 1153   Built in 1979
1154 to 1497   Built in 1980

Tractor Model: 786 B Series

Rated Horsepower   Max 100

Australian Tractor Test #

Engine D-310

Average Shipping Weight: 10,350lbs (4,705kg)

Serial Number Data

2030 to 2251   Built in 1981
2252 to 2441   Built in 1982

Tractor Model: 886

Rated Horsepower   **Max** 121

Australian Tractor Test #

Engine D-358

Average Shipping Weight: 11,700lbs (5,320kg)

Serial Number Data

501 to 675   Built in 1978
676 to 1004   Built in 1979
1005 to 1305   Built in 1980
1306 to 1326   Built in 1981

Tractor Model: 886 B Series

Rated Horsepower   **Max** 121

Australian Tractor Test #

Engine D-358

Average Shipping Weight: 11,700lbs (5,320kg)

Serial Number Data

2001 to 2174   Built in 1981
2175 to 2432   Built in 1982

Tractor Model: 986

Rated Horsepower **Max** 140

Australian Tractor Test #

Engine DT-358

Average Shipping Weight: 12,200lbs (5,545kg)

Serial Number Data

501 to 559	Built in 1978
560 to 817	Built in 1979
818 to 985	Built in 1980
986 to 992	Built in 1981

Tractor Model: 986 B Series

Rated Horsepower   **Max** 140

Australian Tractor Test #

Engine DT-358

Average Shipping Weight: 12,200lbs (5,545kg)

Serial Number Data

2001 to 2155   Built in 1981
2156 to 2328   Built in 1982

Finished September 1982

Tractor Model: 1086

Rated Horsepower   **Max** 148

Australian Tractor Test #

Engine DT-414

Average Shipping Weight: 13,090lbs (5,950kg)

Serial Number Data

501 to 625   Built in 1979
626 to 796   Built in 1980

Tractor Model: 1086 B Series

Rated Horsepower   **Max** 148

Australian Tractor Test #

Engine DT-358

Average Shipping Weight: 12,200lbs (5,545kg)

Serial Number Data

2001 to 2088   Built in 1981
2089 to 2164   Built in 1982

Finished November 1982

Tractor Model: 1486

Rated Horsepower   **Max** 167

Australian Tractor Test #

Engine DT-436

Average Shipping Weight: 13,340lbs (6,065kg)

Serial Number Data

501 to 550   Built in 1979
551 to 570   Built in 1980

Tractor Model: 1486 B Series

Rated Horsepower   **Max** 167

Australian Tractor Test #

Engine DT-436

Average Shipping Weight: 13,340lbs (6,065kg)

Serial Number Data

2001 to 2030   Built in 1981
2031 to 2064   Built in 1982

NOTES

NOTES

NOTES

NOTES

www.ingramcontent.com/pod-product-compliance
Lightning Source LLC
Chambersburg PA
CBHW030535210326
41597CB00014B/1159